ANIMALS OF THE GRASSLANDS

SPEEDY CHEETAHS

THERESA EMMINIZER

PowerKiDS press

New York

Published in 2021 by The Rosen Publishing Group, Inc.
29 East 21st Street, New York, NY 10010

Portions of this work were originally authored by Ethan Grucella and published as *Cheetahs*. All new material in this edition authored by Theresa Emminizer.

Editor: Theresa Emminizer
Book Design: Michael Flynn

Photo Credits: Cover Eugen Haag/Shutterstock.com; p. 5 Freder/E+/Getty Images; p. 7 Paul & Pavveena McKenzie/Stockbyte/Getty Images; p. 9 JonathanC Photography/Shutterstock.com; p. 11 Paul Souders/Stone/Getty Images; p. 13 Bobb Strongman/iStock/Getty Images; p. 15 Krishnan Gopala Krishnan/500Px Plus/Getty Images; p. 17 GUDKOV ANDREY/Shutterstock.com; p. 19 wildatart/iStock/Getty Images; p. 21 Lennart Van Den Berg/EyeEm/Getty Images; p. 22 Eric Isselee/Shutterstock.com.

Library of Congress Cataloging-in-Publication Data

Names: Emminizer, Theresa, author.
Title: Speedy cheetahs / Theresa Emminizer.
Description: New York : PowerKids Press, [2021] | Series: Animals of the grassland | Includes index.
Identifiers: LCCN 2020005894 | ISBN 9781725317123 (library binding) | ISBN 9781725317109 (paperback) | ISBN 9781725317116 (6 pack)
Subjects: LCSH: Cheetah–Juvenile literature.
Classification: LCC QL737.C23 E558 2021 | DDC 599.75/9–dc23

LC record available at https://lccn.loc.gov/2020005894

Manufactured in the United States of America

CPSIA Compliance Information: Batch #CSPK20. For Further Information contact Rosen Publishing, New York, New York at 1-800-237-9932.

CONTENTS

World's Fastest

Cheetahs are the fastest land animals on Earth. They make their homes in the **grasslands** of Africa. Though once widespread, cheetah populations are now dropping. Their **habitats** are shrinking, meaning there are fewer places for them to live and hunt.

Spots and Stripes

Cheetahs have yellow coats covered in black spots. They have black lines on their faces below their eyes, almost like tear tracks. These markings are believed to help keep the sun out of a cheetah's eyes during the day.

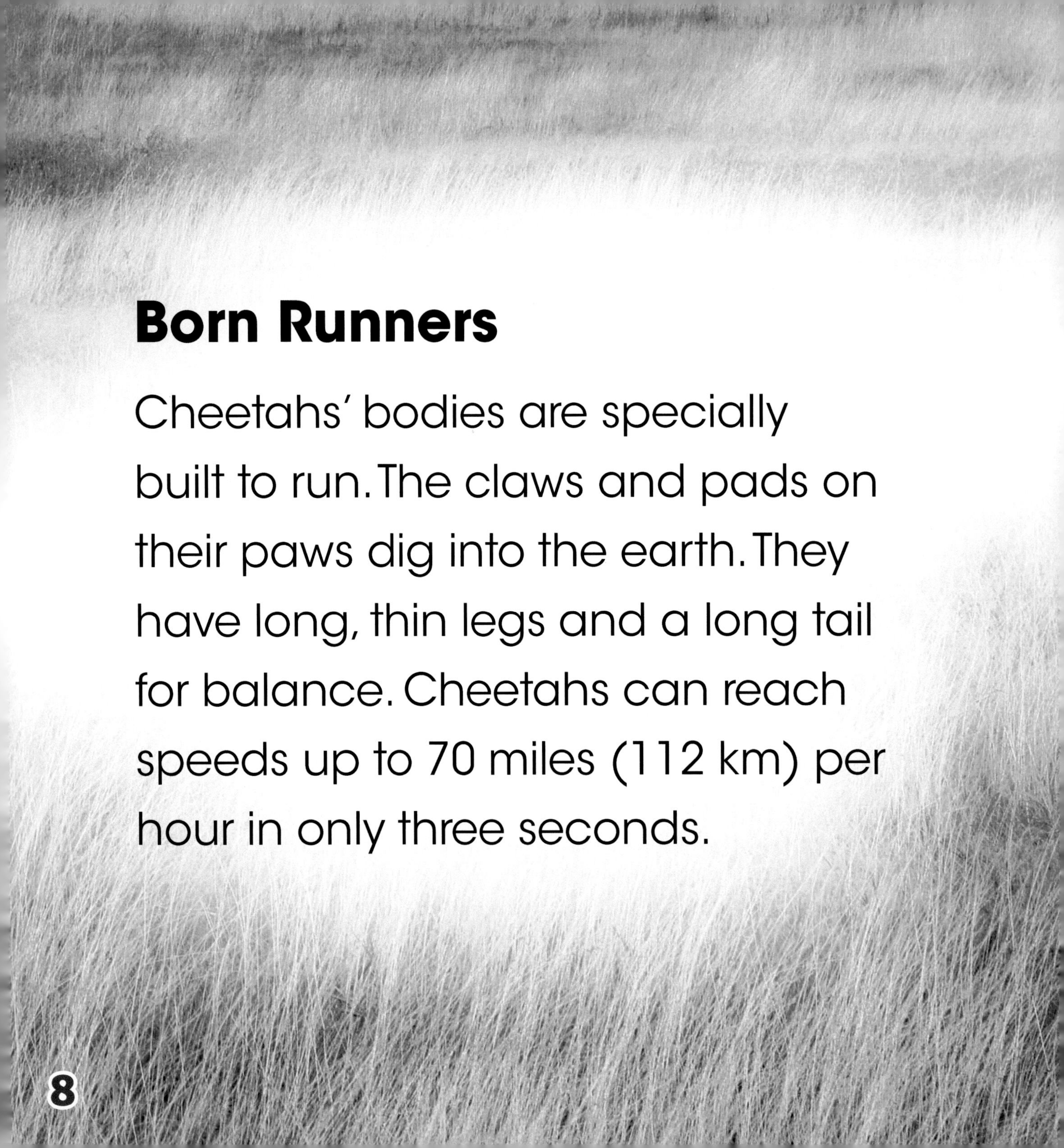

Born Runners

Cheetahs' bodies are specially built to run. The claws and pads on their paws dig into the earth. They have long, thin legs and a long tail for balance. Cheetahs can reach speeds up to 70 miles (112 km) per hour in only three seconds.

Cheetahs run in sprints, or short distances at great speed. A cheetah can cover 20 to 22 feet (6 to 6.7 m) in a single bound. They are the only cats that can turn in midair to follow their prey.

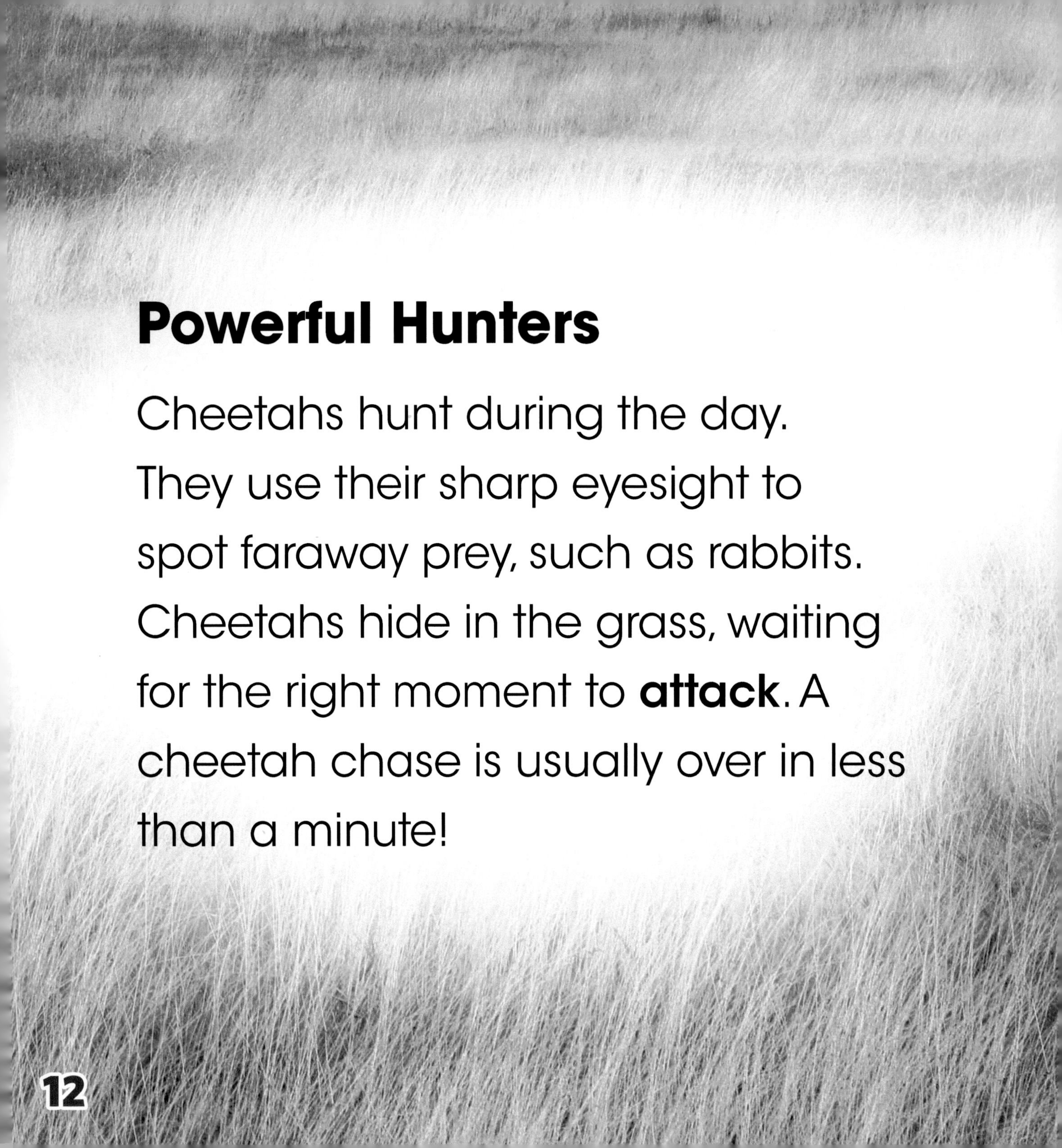

Powerful Hunters

Cheetahs hunt during the day. They use their sharp eyesight to spot faraway prey, such as rabbits. Cheetahs hide in the grass, waiting for the right moment to **attack**. A cheetah chase is usually over in less than a minute!

Once their prey is caught, cheetahs use their powerful jaws and sharp teeth to kill it. Running heats up a cheetah's body to 105°Fahrenheit (41°Celsius). It takes about 20 minutes for the cheetah's body to go back to normal.

Cheetah Cubs

Female, or girl, cheetahs give birth to two to eight cubs at a time. Cubs have a mane on their neck, which goes away as they grow. Cubs stay with their mothers until they're about 18 months old.

A Coalition of Cheetahs

Adult female cheetahs spend most of their time alone, but males, or boys, may live and hunt together. A group of cheetahs is called a coalition. They **communicate** with each other through barks, chirps, purrs, hisses, and calls.

Cheetahs in Trouble

In the wild, cheetahs live for about 12 years. But cheetah habitats are shrinking. Farmers sometimes kill cheetahs. They also suffer from being traded as **exotic** pets. Putting laws in place to protect cheetahs and their homes is more important than ever.

Fancy Features

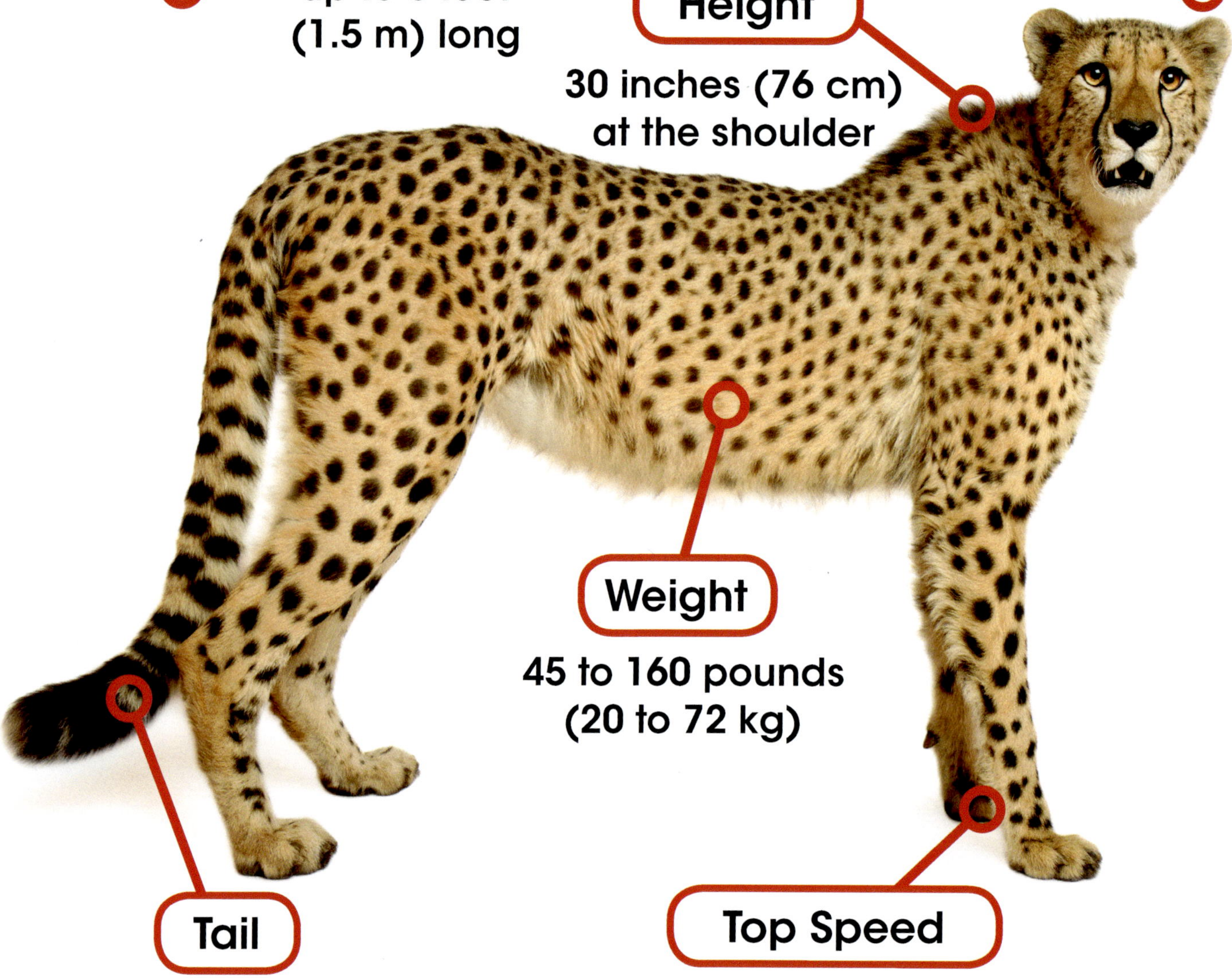

Length

up to 5 feet
(1.5 m) long

Height

30 inches (76 cm)
at the shoulder

Weight

45 to 160 pounds
(20 to 72 kg)

Tail

up to 31 inches
(80 cm)

Top Speed

70 miles (112 km) per hour
in 3 seconds

GLOSSARY

attack: To try to fight with or harm something.

communicate: To talk through sounds and motions.

exotic: Out of the ordinary.

grassland: Land on which grass is the main plant life.

habitat: The natural home for plants, animals, and other living things.

INDEX

WEBSITES

Due to the changing nature of Internet links, PowerKids Press has developed an online list of websites related to the subject of this book. This site is updated regularly. Please use this link to access the list: www.powerkidslinks.com/aotg/cheetahs